BRIDGES

MASTERPIECES
OF ARCHITECTURE

LIONEL BROWNE

This edition published in 1996 by SMITHMARK Publishers,
a division of U.S. Media Holdings, Inc., 16 East 32nd Street, New York, NY 10016.

SMITHMARK books are available for bulk purchase for sales promotion and premium use.
For details write or call the manager of special sales,
SMITHMARK Publishers, 16 East 32nd Street, New York, NY 10016; (212) 532-6600.

This book was designed and produced by Todtri Productions Limited
P.O. Box 572, New York, NY 10116-0572 FAX: (212) 279-1241

Printed and bound in Singapore

Library of Congress Catalog Card Number 96-68168
ISBN 0-7651-9942-4

Author: Lionel Browne

Publisher: Robert M. Tod
Book Designer: Mark Weinberg
Production Coordinator: Heather Weigel
Senior Editor: Edward Douglas
Project Editor: Cynthia Sternau
Editorial Consultant: Hilary Scannell
Assistant Editor: Linda Greer
Picture Researcher: Natalie Goldstein
Desktop Associate: Michael Walther
Typesetting: Command-O, NYC

CONTENTS

INTRODUCTION

Bridges have always exerted a hold on the imagination; perhaps in no other form of structure is technology so closely intertwined with aesthetics. From the breathtaking boldness of a Roman aqueduct, to the enduring vigor of the Rialto Bridge in Venice, to the soaring gossamer elegance of a modern suspension bridge, the history and development of bridge design and bridge-building over the centuries have paralleled man's increasing understanding of the strength of materials and of the interplay of forces in structures made from those materials.

The first bridge constructions were based on naturally occurring materials—usually wood or stone. As soon as the first hunter-gatherer tribes began to range further afield for food, fuel, and shelter, they came across natural obstacles: streams, rivers, chasms, or gorges. Sometimes they were able to take advantage of a natural rock span, created by running water as it ate away soft strata from under harder layers of rock, but such rock formations were rare. More probably our remote ancestors would have taken advantage of dead tree trunks or branches that had toppled across gaps, or would have used a hanging jungle vine to swing across a stream. As tools were developed, it became possible to cut down a tree trunk and lay it across a gap. In mountainous terrain, rocks would have been used, either singly or propped together in a primitive arch.

Every bridge, even the most primitive, uses the properties of its materials to overcome and exploit the forces that it is subjected to. There are three basic sources of the forces on a bridge, called loads. Dead load is the bridge's own weight; live load is created by the traffic passing over it; and environmental load is imposed by external forces, such as wind, water, or earthquakes.

The pattern of forces created by these loads is often extremely complex, but engineers classify the forces themselves into basic types. Tension pulls apart, and compression—its opposite—pushes together. Shear is a cutting or shearing force, and torsion is a twisting force. The various strengths of a material, such as tensile strength and compressive strength, measure its ability to resist these four forces. Stone has considerable compressive strength, but not much tensile strength, while vines and rope have tensile strength, but no compressive strength. Different types of wood combine tensile and compressive strength in different proportions. A solid oak tree trunk has nearly as much compressive strength as stone, but considerably more tensile strength. However, like all natural materials, it is what an engineer would call weak in shear—it can snap. So too, obviously, can vines and ropes, but they are much better able to resist torsion (twisting) than either timber or stone. Iron and steel are superior to these natural materials in all four strengths, and reinforced concrete combines the high tensile strength of steel with the compressive strength of mass concrete.

RIGHT: The Ponte Sant'Angelo in Rome was built in A.D. 134. Originally known as the Pons Aelius, it consists of five small arches spanning the River Tiber. The present balustrade, including a cast-iron railing, was added in the seventeeth century by Bernini, as well as the statues of angels above the piers.

ABOVE: The huge towers of the Golden Gate Bridge are 746 feet
(227 meters) high. The cables, which weigh 7,125 tons (6,412 metric tons)
apiece, were each spun from over 27,000 wires. Work began on the
bridge in January 1933, and it was first opened to traffic in May 1937.

The simplest type of bridge—a log or stone across a stream—is technically known as a beam. The combination of its own weight (dead load) and traffic across it (live load) will tend to make it bend. The result will be tension along its top surface and compression along the bottom. In the beam, the compressive and tensile forces are balanced, while the overall mass of the bridge bears directly downwards on the ground beneath.

The cantilever is a development of the beam. It consists essentially of a bracket with one end fixed and the other end hanging free. The simplest cantilever bridge has a beam supported on two opposite cantilevers. More typically a cantilever bridge will have a number of spans, with the cantilevers balanced evenly from piers. A cantilever gener- ates some force outwards in addition to the direct down- ward force.

The simple corbeled arch—an intermediate form between cantilever and true arch—was known to the Sumerians and Egyptians nearly three thousand years ago. It consists of suc- cessive courses of masonry placed on either side of an open- ing, projecting closer and closer together until they meet.

The true arch form was also known in Egypt and Mesopotamia, almost as early as corbeling. It is constructed of wedges of stone, called voussoirs, placed in a semicircle. Dead and live load combine to force the material of the arch togeth- er in compression. Because the arch is in compression only, stone makes an ideal construction material, as it is not sub- jected to any tensile force.

ABOVE: A backpacking tourist cautiously navigates fallen trees lying
across a swollen mountain stream. Our early ancestors would have
been familiar with bridges very much like this, in common use before
the invention of cement and the development of structural engineering.

CHAPTER ONE

USING WHAT NATURE PROVIDED: STONE AND WOOD

The first bridges would have been built to help hunter-gatherer tribes range further afield. But about ten thousand years ago the rise of agriculture saw the growth of communities, first into villages, and then into city-states. The infrastructure that these required—buildings, irrigation, and so on—led to the development of engineering skills.

The Ancient World

In the seventh century B.C., Sennacherib built a network of canals around Nineveh (capital of the Assyrian Empire, now in modern Iraq). This included a stone aqueduct 920 feet (280 meters) long and 66 feet (20 meters) wide across a small river valley at Jerwan. Under the rule of Nabopolassar and his son Nebuchadnezzar II, Babylon (also in modern Iraq) became

LEFT: The Kintai-Kyo Bridge, Iwakuni, Japan, was originally erected over the Nishiki River in 1673. The entire timber structure was then renewed four times in each century by rebuilding one of the five arches every five years. Finally, the complete bridge was rebuilt in 1953, using modern preservative chemicals to treat the timbers.

the greatest metropolis the world had yet seen. A dominant feature of the city was a bridge across the Euphrates River, close to the stepped tower that is now identified with the biblical Tower of Babel. This bridge is thought to have been between 390 and 650 feet long (120–200 meters); excavations have uncovered the remains of seven piers in baked brick, each roughly 30 by 65 feet (9 by 20 meters).

The Romans were the first to develop real expertise in the design and construction of bridges. Most Roman bridges were probably timber, not stone, but in general only the stone ones have survived. (Julius Caesar records a timber trestle bridge perhaps a quarter of a mile [400 meters] long, which he ordered to be built across the Rhine in 55 B.C.) By the second century B.C., the Roman engineers had mastered the techniques of creating secure midstream foundations and constructing masonry works above. Some of their bridges have survived as usable structures for two thousand years. How were the Romans able to build such durable structures? There were three key factors. First, they developed an excellent waterproof cement called *pozzolana*, made by mixing water,

lime, and sand with fine powder ground from volcanic rock found near Pozzuoli (hence the name). Second, they developed a method of constructing foundations within a temporary enclosure called a cofferdam. And finally, they realized the potential of the voussoir arch, which could span greater distances than any unsupported beam, and was stronger, more secure, and more durable than any other structure that could be built with the materials available at that time.

Roman arches were semicircular, so much of the loading was directed vertically downward. Hence the Romans could build their bridges out from the shore one span at a time (first cofferdam, then foundations, then pier, then arch), rather than having to try and build all the foundations at once. The massive piers needed to support the semicircular arches reduced the width of the channel, and speeded up and concentrated the flow of the current.

Arches were constructed using a wooden formwork, or centering. This was built out from the piers, with the upper surface shaped to the required profile. Parallel arcs of stones were then placed behind each other to form the arch. The semicircular shape meant that all the stones were cut identically, and no mortar was necessary. Once the keystone in the center was positioned, the compressive forces, together with the perfect cut and fit of the blocks, ensured complete stability.

Aqueducts

The Roman Empire had an extensive network of aqueducts; Rome alone had eleven, the last of which was completed in A.D. 226. The two most impressive to have survived are the Pont du

Gard, near Nîmes in southern France, and the aqueduct at Segovia in central Spain, which was built at the beginning of the second century A.D. to bring water to Segovia from the Guadarrama Mountains 62 miles (100 kilometers) away. It is 2,624 feet (800 meters) long, with a double row of arches, one above the other, reaching to 118 feet (36 meters) at the highest point.

The Pont du Gard was probably built by the Emperor Augustus as part of a 25-mile (40-kilometer) aqueduct. It is 886 feet (270 meters) long and spans the River Gard at a maximum height of 160 feet (49 meters). It has three tiers of arches: two broad spans of symmetrical arches—six below and eleven above—and a top row of thirty-five smaller arches, which originally supported a deep, cement-lined channel that carried the water. The main arches are extremely broad: the longest, across the river, is 80 feet, 6 inches (24.5 meters). In Roman bridges the ratio of pier to span widths is usually 3:1, but in the Pont du Gard it is more like 5:1. Although the Pont

ABOVE: An abandoned gypsy wagon rests in front of the Doppelbrücke (the Double Bridge), in Rothenburg, Germany. The Doppelbrücke dates back to ancient Roman times, when the invention of cement first made the construction of masonry bridges possible.

du Gard has suffered much damage over the years, its soaring arches have nevertheless stood for two thousand years—without the benefit of any mortar.

Many notable masonry bridges were built in Persia and China before the twelfth century A.D. In A.D. 260 the Persian ruler Shapor I used Roman captives to build a bridge over the Karun River. With forty arches, it had a total length of 1,640 feet (500 meters). The form of the semicircular arch also influenced Chinese bridge-building, but the An Ji Bridge at Zhao Xian in Hebei Province was revolutionary in its design. Built in A.D. 605, it had a single span of 121 feet (37 meters). Instead of being a complete semicircle, the arch was segmental: the arc forming the span was only a small part of a circle. The haunches on either side were also pierced by pairs of smaller arches, to reduce the weight of the masonry and hence to lighten the load. The result was an elegant span that rose to a height of only 23 feet (7 meters); nothing comparable was to be built in the West for another seven hundred years.

ABOVE: This ancient Celtic stone bridge in the Scottish highlands is a graphic demonstration of the loadbearing capacity of the arch form. The slender semicircle of voussoirs transmits the loading of the bridge downward and outward to the far more massive abutments.

Masonry Arch Bridges

The semicircular Roman arch continued as a tradition in church-building until the middle ages, when it was superseded by the pointed arch. This was easier to build than the semicircular arch, as the voussoirs did not have to be matched and dressed as precisely.

The year 1209 saw the completion of Old London Bridge. This most famous of medieval bridges had been begun thirty years previously by Peter of Colechurch, the Chaplain of St. Mary's Church, Cheapside. It had nineteen small pointed arches of irregular sizes spanning equally irregular boat-shaped piers or starlings. This was the first stone bridge with masonry foundations to be built in a swiftly flowing tidal river. The starlings were so wide that the total waterway was reduced to a quarter of its total width. Upstream of the bridge the water moved so slowly that it froze over in severe winters. The level of the river water fell by up to 4 feet (1.2 meters) as it came through the arches, and downstream of the bridge the force of the current scoured out the bed of the river to form a

BELOW: The Rialto Bridge has safely spanned the Grand Canal in Venice since its completion in July of 1591. Designed by the engineer Antonio da Ponte— by then in his seventies—the single 88.5-foot (27-meter) stone and brick arch is supported on thousands of wooden piles driven into the soft ground.

LEFT: The Pont Valentré at Cahors, France, is a remarkably well-preserved example of a medieval fortified bridge. Bridges such as this were defended by ramparts, towers with firing slits, and often a drawbridge.

ABOVE: Old London Bridge, painted about 1750 by J. Varley (1778–1842) clearly shows the force of the current passing through the narrow arches set over the Thames.

deep pool capable of taking the largest ships. This stretch of the river is still known as the Pool of London.

Despite the relative crudity of its construction, Old London Bridge survived until 1831. In 1763 it was stripped of its overtopping street of timber-framed houses, and had two of its central arches replaced by a new navigation arch. Old London Bridge was not unique in having structures on it; notable later bridges to follow suit were the Ponte Vecchio in Florence, the Rialto in Venice, and the Pulteney Bridge in Bath, England.

Exactly contemporary with the original London Bridge was the Pont d'Avignon over the Rhône in France. The bridge at Avignon had one or two more spans than London Bridge, but was three times as long, spanning a total of 3,000 feet (914 meters) across the Petit-Rhône, over the island of Barthelasse and turning in the middle of the Grand-Rhône in a V shape, which was intended to withstand the high spring floods. Only four spans have survived, but they are longer than any Roman arch and are elliptical in form, resulting in narrower piers and taller spans.

The Pont d'Avignon was a tremendous engineering achievement. So too, in the next century, was the Ponte Vecchio in Florence, the first European use of the segmental arch—a portion of a circle, rather than a full semicircle—a design already been seen in China at the An Ji Bridge some seven hundred years earlier.

ABOVE: The Pont d'Avignon in France was justifiably one of the most famous of medieval bridges. Completed in 1188, its elliptical arches spanned 3,000 feet (900 meters) across the River Rhône. Today all that remains is these four arches, bearing the little chapel of St. Bénézet, reputedly the bridge's designer.

RIGHT: A landmark masonry bridge over the Rhein River in Worms. One of the most venerable historic centers of Europe, this German city was originally a Celtic settlement called Borbetomagus.

LEFT: This view of the River Seine from Paris' Eiffel Tower shows—from left to right—the Pont de l'Alma, the Pont des Invalides, the Pont Alexandre III, and, in the distance, the Pont de la Concorde.

BELOW: In 1578 work was begun on the Pont Neuf in Paris, and the bridge was completed in 1607. Its twelve arches—all of different spans—were flattened from semicircular to elliptical during reconstruction work in the nineteenth century.

LEFT: A gondola slips quietly beneath the Bridge of Sighs (Ponte dei Sospiri) in Venice. The bridge, which was designed by Antonio Contino in 1560, links the Doge's Palace with the prisons. It is a romantic feature, with an elliptical arch, rusticated pilasters, and heraldic devices, which was used for the passage of prisoners to be examined by the Inquisitors of the State.

The Ponte Vecchio

The Ponte Vecchio that stands today had two predecessors, both destroyed by floods. Work began on the new bridge in 1345; the design of the bridge is attributed to Taddeo Gaddi, a pupil of the painter Giotto. The three well-balanced arches are remarkably shallow in profile, with a rise that varies from 12 feet, 9 inches (3.9 meters) to 14 feet, 6 inches (4.4 meters). The consequent low height of the bridge deck resulted in a shallow roadway which was wide enough to allow the construction of two stories of buildings along the bridge. Today these house a double row of jewelers' shops, surmounted by a gallery linking the Pitti and Uffizi palaces.

The spans of the Ponte Vecchio were relatively modest, varying between 90 and 100 feet (27 and 30 meters), but within a few years the segmental arch had been developed further. The Ponte Castelvecchio, completed in 1356, stretched nearly 164 feet (50 meters) across the Adige River at Verona; the year 1371 saw the completion of a bridge with a 236-foot (72-meter) span, built by the Duke of Milan across the River Adda at Trezzo. This arch was over twice the span of anything built by the Romans, and set a record that was not beaten until the eighteenth century.

ABOVE: This distant view of the Ponte Vecchio, Florence, clearly shows the form of the arches, daringly shallow in their time. Work began on the bridge in 1345; until then the form of stone arches had generally followed the Roman semicircular model. The Ponte Vecchio has segmental arches—these are lower, wider, and therefore able to span greater widths than semicircular arches.

Less innovative, but much grander in scale, was the Karlsbrücke in Prague. Originally there had been a wooden bridge over the Vltava, but it was destroyed by a flood in the middle of the twelfth century and replaced by a stone bridge known as the Judith Bridge, named after the wife of King Vladislav I. This bridge stood for 170 years, but was destroyed by another flood in 1342.

The present Karlsbrücke was begun in 1357 by Charles IV (crowned Holy Roman Emperor in 1346). It was in use by the 1380s, but successive floods weakened the structure, which was built in sandstone rather than in much harder granite. The most serious damage occurred in 1432, when the bridge was weakened in three places.

The bridge is carried in a slight zigzag on sixteen arches across a total span of 1,692 feet (516 meters), the longest con-

RIGHT: The piers of the Ponte Vecchio—made suitably massive to resist the horizontal thrust generated by the shallow arches—were later equally able to support this superstructure. An upper gallery (just visible at the left) connects the Pitti and Uffizi Palaces; on the lower level is a double row of jewelers' shops.

BELOW: Tourists throng the Karlsbrücke (Charles Bridge) over the River Vltava (Danube) in Prague. Visible along either side of the bridge are the statues—thirty in all—that have been added to the bridge over a two-hundred-year period, from 1683 to 1859.

tinuous stretch of water covered by any medieval bridge. It is an amalgam of different styles: Roman-style arches are carried on typically medieval piers; and the lower tower on the left bank dates back to the twelfth century (a relic of the old Judith Bridge), but was renovated in 1590 in Renaissance style; the higher towers on each bank were built in matching styles in the fourteenth and fifteenth century.

The Renaissance and Beyond

Bridge design and bridge-building flourished in the Renaissance. In 1502, Leonardo da Vinci produced a design for a masonry arch bridge with a clear span of 787 feet (240 meters) over the Golden Horn at Istanbul. This was not built, but many less grandiose bridges were, including the Pont Notre Dame and Pont Neuf in Paris, the Rialto in Venice, and the Ponte Santa Trinità in Florence. The last, built from 1567 to 1569, displays a remarkable development of the curve in what, despite its elegance, is now known somewhat prosaically as a basket-handled design. The three arches are nearly elliptical; they are extremely shallow at the crown, and concealed behind decorative pendants, but spring vertically from the piers. The bridge was destroyed in the Second World War, but was subsequently rebuilt in its original form, using many of the original materials.

Bartolomeo Ammanati, who designed the Ponta Santa Trinità, based the shape of the curve on scientific principles, and it was during the Renaissance and on into the eighteenth centu-

BELOW: This classic Amsterdam bridge was designed by H. P. Berlage (1856–1934), a Dutch architect who took part in city planning projects for the Hague and Amsterdam. His ideas won great favor with the rising generation of modern architects, including the Amsterdam school.

ry that the study of structures and the forces acting on them began to pay increasing dividends in the design of bridges.

In 1720, the Corps des Ingénieurs des Ponts et Chaussées was founded, and in 1747, Daniel-Charles Trudaine opened the Ecole des Ponts et Chaussées, the first engineering school. The school's director for nearly fifty years was Jean Perronet, who was the most famous bridge-builder of his day. Perronet applied the latest advances in structural analysis and strength of materials to his designs, and showed that traditional forms of construction were unnecessarily conservative. As a result, his arches became ever shallower and thinner. The Pont de Neuilly featured a development of the *corne de vache* ("cow's horn") design, in which flattened elliptical curves were shaved off to become shallow segmental arch profiles of the facades. This device was combined with extremely slender piers to yield (to modern eyes) a refreshingly elegant and uncluttered appearance. The Pont Saint-Maxence went even further; it had a vertical distance from springing to crown of only one twelfth of the span, a value seldom exceeded since.

In 1810, the Strand Bridge Company commissioned John Rennie to design a bridge over the Thames in London, after he had roundly criticized a design based on Perronet's Pont de Neuilly: instead of using the *cornes de vache* design, he insisted that the bridge "should either be a plain ellipse . . . or . . . the flat segment of a circle." And so it was that the first Waterloo Bridge (opened in 1817) combined Rennie's plain ellipses with pairs of Doric supports that rose from the piers to projecting refuges on the parapets, producing an effect reminiscent of a row of classical temples.

By 1820 it was clear that, although Old London Bridge had stood for seven hundred years, its days were numbered. The rows of houses had been removed in 1763, but the roadway was too narrow to cope with the increase in traffic. The starlings and piers were constantly being damaged by boats and needed repair. The Corporation of London called in John Rennie. His verdict was that the bridge should be replaced, and he produced a design with five semi-elliptical arches. However, Rennie died in 1821 and the building of his design was entrusted to his son. New London Bridge was opened in 1831. Replaced in 1972, it now stands reassembled in Lake Havasu City, Arizona.

Timber

When man first began to build bridges, tree trunks were the most readily available material for bridge construction. Multiple raked frames constructed by the Romans achieved spans of up to 100 feet (30 meters). Notable examples were the Rhine bridges at Mainz in the first century A.D., and at Cologne in the fourth century A.D. The art of timber construction flowered in the middle ages; bridges were constructed with deck-stiffening raked frames or built-up timber arch spans of up to 200 feet (60 meters).

In the Renaissance, Andrea Palladio described several types of timber truss bridge in his definitive work *I Quattro Libri dell' Architettura* (*The Four Books of Architecture*). In the general sense, a truss is a structural framework that can support weight. It exploits the rigidity of the triangle and the balance of tensile and compressive forces to achieve the same structural result as a beam or an arch, but with dramatically less material and hence less weight.

RIGHT: Amsterdam's Magere Bridge. Sometimes called the "Venice of the North," Amsterdam is transversed by about forty canals; these are flanked by streets and crossed by some four hundred bridges.

ABOVE: The nearer bridge in this picture of the River Vltava (Moldau) in Prague, is the May Day Bridge. It was built in 1901 to replace a nineteenth-century suspension bridge, which was the first to span the river after the fourteenth-century Karlsbrücke, here next in view.

LEFT: The Schlossbrücke, crossing the Schleusenkanal at the end of Unter den Linden in Berlin, leads to the historic Zeughaus (armory). Finished in 1824, this bridge bears martial statues celebrating the war of liberation, which were carved by pupils of Christian Rauch in the mid-1850s.

FOLLOWING PAGE: The historic Theodor Heuss Bridge over the Neckar provides a splendid view of the famous ruined castle which looms over the university town of Heidelberg.

In seventeenth-century England a type of bridge was developed in which the struts of the superstructure were arranged in a strict mathematical order. Canaletto's painting *Old Walton Bridge* shows a structure of this type. Timber bridge-building also flourished in the Far East at this time, notably in Japan.

The Swiss carpenter brothers Johannes and Hans Ulrich Grubenmann (1701–1771 and 1709–1783, respectively) continued the tradition of timber bridge-building into the eighteenth century. Their bridge over the Limmat River near Wettingen spanned 200 feet (61 meters). In 1755, they were commissioned to rebuild the bridge over the Rhine in Schaffhausen—a span of 390 feet (119 meters). Their model of the bridge, 16 feet (5 meters), still survives, and can be seen in the city museum.

In 1785 Colonel Enoch Hale designed and built a bridge to carry a turnpike forming part of the main trading route from Boston to Montreal across the Connecticut River at Bellows Falls in Vermont. Such records as survive suggest that the bridge was between 300 and 400 feet long (90–120 meters), supported 50 feet (15 meters) above the river by a central tim-

ABOVE: The Kapellenbrücke spans the Vierwaldstätter See, a large lake in Lucerne, Switzerland. Built at the beginning of the fourteenth century, its basic structure consists of simple timber beams laid on closely spaced piles, braced and raked together. The roofed form—a sensible precaution against the ravages of the weather—was to be echoed centuries later in the covered wooden bridges of nineteenth-century America.

RIGHT: The Cornish–Windsor Bridge between Vermont and New Hampshire is a splendid example of the "creaking, dark, mossy tunnels"—the covered bridges—that characterized rural, pre-industrial America.

ABOVE: As the American railroad network spread across the country, the abundant pine forests provided the raw material for the hundreds of bridges that were needed—even well into the twentieth century. This wooden railroad bridge in Carriso Gorge, California, was built in 1921; it is 185 feet (56 meters) high and 633 feet (193 meters) long.

ber pier, and braced with four sets of inclined struts. The bridge survived in modified form until 1840.

Covered bridges seem to have made their first appearance in 1805, when Timothy Palmer erected a triple-span arch over the Schuylkill River in Philadelphia. The president of the Schuylkill Bridge Company insisted that the bridge should be completely enclosed in boarding to protect it.

Palmer was one of the first professional bridge-builders in America. Another was Lewis Wernwag, who was a pioneer of timber cantilevering. His best-known work was the Colossus Bridge, also built over the Schuylkill River, at Fairmount Park, Philadelphia. This had a single span of 340 feet (104 meters). Theodore Burr exceeded this with his 360-foot (110-meter) timber arch over the Susquehanna at McCalls Ferry, southwest of Lancaster, Pennsylvania. Sadly, this was destroyed by ice two years later, but hundreds of Burr's other bridges are still in use.

With the arrival of the railways on both sides of the Atlantic, there was a pressing need for bridges that could be built quick-

ly. In Britain, the great engineer Isambard Kingdom Brunel (1806–1859) was foremost in expanding the railways. Between 1849 and 1864 he constructed sixty-four viaducts for the South Devon, West Cornwall, and Cornwall lines. All were made of timber.

In America, ten years after Long's patent, an architect named Willam Howe from Massachusetts patented a very similar design, but with one vital difference: the vertical members were made not of wood, but of iron. And it is to the development of iron bridges that we now turn.

CHAPTER TWO

THE NEW MATERIALS: IRON, STEEL, AND CONCRETE

Although Robert Mylne produced a design for a small, two-arch iron bridge at Inverary, Strathclyde, in Scotland, it was never built. The honors for producing the first cast-iron arch span must go to Thomas Pritchard. This was a semicircular arch with a 100-foot (30-meter) span over the River at Coalbrookdale, Shropshire, in England. The bridge took a mere three months to erect. The members were cast in local blast furnaces; the five main ribs were cast in two halves, each weighing nearly 6 tons. Cast iron is stronger in tension than masonry, but its main advantage is its much greater compressive strength; yet despite the relative slenderness of the ribs and connectors of the Coalbrookdale arch, it is "over designed," with much redundant cast iron in the structure.

The Coalbrookdale Bridge used the traditional semi-circular arch, while its jointing techniques were very much in the tradition of timber construction, with its mortise joints and dovetails. But as the Industrial Revolution got into its stride in the nine-

teenth century, the techniques of bridge-building in iron—and later, in steel—were to develop dramatically.

The next developments in the use of iron for bridges came from the revolutionary political philosopher Thomas Paine (1737–1809). In 1788 he took out a patent for a new type of iron bridge. Segmental in shape, it had five cast-iron ribs and a span of 110 feet (34 meters). Thomas Wilson (1736–1819), an engineer who had served as agent to James Watt, built a

LEFT: The New River Gorge Bridge, built in 1978 in West Virginia, is the world's longest steel arch bridge, with a span of 1,700 feet (518 meters), and a total length of 3,030 feet (924 meters). Soaring 867 feet (267 meters) above the river below, it is nearly the world's highest span as well.

ABOVE: The Iron Bridge at Coalbrookdale in Shropshire, England, the world's first cast-iron bridge, has become an icon of the Industrial Revolution, and is now the centerpiece of a national museum.

ABOVE: This unique view of Waterloo Bridge, London, taken in 1924 with a telephoto lens, shows strengthening work in progress on the foundations of the bridge. At that time, the structural situation was considered to be serious, and prompt action was taken to correct it.

variation of Paine's design over the River Wear at Sunderland in the north of England between 1793 and 1796. Like the Coalbrookdale Bridge, the Sunderland Bridge had five iron ribs built up of cast-iron panels, but its span was more than double, at 236 feet (72 meters), with a rise of 34 feet (10 meters).

The Great Engineers

The next developments in bridge-building in iron were the work of the two great engineers Thomas Telford and John Rennie. Telford was appointed Surveyor of Public Works for Shropshire in 1788, and he designed forty-two bridges in England, some in stone and some in iron. His first iron bridge was built over the Severn at Buildwas in 1795, not far from Coalbrookdale. However, like the Sunderland Bridge, it was segmental in form rather than semicircular, and although it

was 30 feet (9 meters) longer than the Coalbrookdale Bridge, it used only 173 tons of iron, compared with Coalbrookdale's 378 tons.

The development of iron bridges was not all clear sailing. Thomas Wilson, who built the Sunderland Bridge, built a cast-iron bridge over the River Thames at Staines in 1803, but it had to be demolished later the same year when the ironwork began to fracture. An iron bridge built at Bristol in 1805 by William Jessop (1745–1814), with castings from the foundry at Coalbrookdale, collapsed soon after completion.

Despite such setbacks, confidence in the use of iron continued to grow. After Telford had been appointed as engineer to Scotland's Highland Road Commission in 1803, he built a number of stone bridges, but resorted to iron when a particularly large span was required. But the honors for the largest cast-iron arch to be built in Britain go to Rennie. In 1800, he had built his first iron bridge, at Boston in Lincolnshire. It was modest in scale, but notable for being built in wrought iron rather than cast iron. Wrought iron was first made in large quantities in the 1780s by Henry Cort, using his puddling process. It has a lower carbon content than cast iron, making it softer and more malleable; it has less compressive strength than cast iron, but a higher tensile strength. The bridge at Boston weighed just over 3 tons. One bridge historian suggests that a comparable bridge in cast iron would have weighed 208 tons!

Rennie designed three toll bridges over the Thames in the space of ten years. Vauxhall Bridge and Waterloo Bridges were both to have been built in stone, but Rennie's design for Vauxhall Bridge was considered too expensive, and instead the directors of the toll company commissioned a nine-arch iron bridge from the naval engineer Sir Samuel Bennett (1757–1831). However, Rennie's Waterloo Bridge *was* built, and so was the third bridge, over the Thames at Southwark.

The site was a deep, narrow point on the river, which could be reached by ships if Old London Bridge were to be rebuilt (which of course it was, in 1831). The three segmental arches at Southwark were of iron; the center arch had a span of 240 feet (73 meters), rose 24 feet (7 meters) above high-water level, and weighed 1,300 tons. This huge weight was supported on massive stone piers, made of huge granite blocks weighing up to 25 tons, which had been quarried in Scotland.

Tubes, Trusses, and Cantilevers

When Robert Stephenson designed the Britannia Bridge to carry his Chester & Holyhead railway across the Menai Strait in north Wales, he had to resort to model testing to arrive at the final design. His chosen site was about a mile (1.6 kilometers) south of Thomas Telford's road suspension bridge, where a tiny midstream island could support a central pier. The

BELOW: During the Cold War, no one from the West was permitted to cross into East Germany by means of Berlin's Glienicker Bridge—used almost exclusively at that time for the exchange of spies between the East and the West.

Admiralty insisted that the waterway should not be obstruct-ed or reduced, so the two main spans of the bridge had to be 460 feet (140 meters). Suspended decks of this width would have been too vulnerable, and solid beams would have had to have been so deep for rigidity that they would have been impossibly heavy. The only practical solution was wrought-iron tubes. Theory suggested that these would need to be sup-ported by chains. However, Scottish ship-building engineer Sir William Fairbairn carried out tests that demonstrated conclu-sively that a rectangular-section tube could stand without additional support. Even so, the towers on the bridge were built high enough to carry chains if required, and were equipped with slots to accommodate them.

The wrought-iron plates and sections were fabricated on site, and the side sections of the tubes were riveted together partly by hand and partly using hydraulic machines designed by Fairbairn. However, the main sections, each 472 feet (144 meters) long, had to be prefabricated, and then floated out on pontoons and raised to their final level by means of huge hydraulic jacks on the piers.

The use of iron for bridges was developing elsewhere in the world. In America, other designs followed William Howe's patent, meeting the need of the railroad companies for a quick, economical way of carrying their expanding networks over the numerous obstacles in their path. The first major iron-truss bridge was built in the United States in 1851, and the earliest iron cantilever girder bridge was built over the Main River at Hassfurt, Germany, in 1867.

Although in 1847 Squire Whipple in the United States had published a treatise on bridge-building, using scientific calcu-lations, much was still to be learned about the properties of materials. In December 1876, an eleven-car train hauled by two locomotives plunged from a bridge at Ashtabula, Ohio, killing many of those on board. This had been the first Howe truss bridge to be made entirely from wrought iron. Other failures followed, and it was not until the use of steel became common that truss bridges could reliably be called safe.

Steel

Cast iron contains a relatively high proportion of carbon—between two and four percent. Wrought iron has most of the carbon removed. In the various types of steel, a controlled amount of carbon is replaced, together with small amounts of other materials such as chromium, nickel, and manganese. Steel combines most of the advantages of cast and wrought iron without the disadvantages, but it could not be used as a bridge-building material until industrial techniques enabled it to be produced in sufficient quantity.

The St. Louis Bridge across the Mississippi, completed in 1874, was the first really big bridge to be built of steel, using chrome steel patented by Julius Barr. The designer of the bridge, James B. Eads (1820–1887), had been appointed engi-neer-in-chief of the newly formed bridge company. He had spent twenty years designing various vessels, including a fleet of ironclad gunboats for the Union during the Civil War, but this was to be his only bridge.

LEFT: The city skyline is visible under the Greater New Orleans Bridge transversing the Mississippi River. This cantilever design was completed in 1958. Its span is 1,575 feet long (480 meters).

RIGHT: New London Bridge, designed by John Rennie, was opened in 1831. It survived until it was replaced by today's concrete structure, and now stands reassembled in Lake Havasu City, Arizona, as a tourist attraction.

ABOVE: A highway land bridge leads to the historic hilltop town of Eze, in the French Riviera. These modern concrete arches are reminiscent in form of those used by the ancient Romans in the construction of the Pont du Gard, near Nîmes.

RIGHT: An aerial view of the ceremony to open the new London Bridge on March 16, 1973. Tower Bridge can be seen in the background and, to its left, the Tower of London.

He designed it with three steel arches with spans of 502, 520, and 502 feet (153, 158, and 153 meters). It was made double-decked, with a roadway above and a railroad below. The Mississippi has a deep and treacherously shifting bed of sand. To reach bedrock, the foundations for the bridge had to be excavated more than 100 feet (30 meters) deep. For the men sinking the foundations, this meant working under compressed air at greater depths than ever before. Little was known about working under these conditions, and the need for slow decompression; of the six hundred men on site, one hundred and nineteen suffered serious cases of the bends and fourteen died.

Eads' bridge, which set a new standard for bridge design and specification in steel, was nevertheless traditional in using masonry piers. The first all-steel bridge was erected in 1879 by General William Sooy Smith. He used a new type of steel developed by A. T. Hay to build five 311-foot (95-meter) trusses to carry a railroad across the river at Glasgow, Missouri.

Problems and Solutions

There were setbacks to the advance of steel as a replacement for wrought iron, particularly where the structural analysis had been inadequate. Most memorable, perhaps, is the collapse of the great railway bridge over the Firth of Tay, near Dundee, Scotland on December 28, 1879. The central thirteen spans, known as the "High Girders," had no braced connection with the bridge on either side. As a mail train crossed the bridge at 7 PM, in a violent storm, the High Girders were blown down, together with cast-iron columns, the train, and seventy-five passengers. It was the country's worst-ever railway accident. The designer of the bridge was the engineer Sir Thomas Bouch. He was described by the ensuing Court of Inquiry as having been "fatally complacent"— particularly about the effect of wind pressure, which he had scarcely considered.

The Firth of Forth, west of Edinburgh, is narrower than the Tay at Dundee, but deeper. There is only one small midstream

ABOVE: When it was completed in 1889, the cantilevered Forth Railway
Bridge, just west of Edinburgh, broke all records for sheer size and
strength. Yet despite its bulk, it displays a remarkable grace and elegance.

island to permit the construction of an intermediate pier. After the Tay disaster, Bouch's design for a Forth crossing was abandoned. With the insight gained from the Tay collapse, the chosen designers—Sir John Fowler and Benjamin Baker—were required to design a structure that could withstand wind pressures of up to 56 pounds per square foot. To achieve this, they combined the use of steel with the cantilever principle, to produce a bridge that broke all records: the volume of masonry in the piers, the dimensions of the cantilevers, the free spans, and the 58,000 tons of steel were all greater than anything seen before. Not only was it the biggest bridge in the world, it was also the strongest and the stiffest. Each of the two main spans consisted of two 680-foot (207-meter) cantilever arms, with a 350-foot (107-meter) suspended truss between them. The spans were built out as balanced cantilevers from each main pier, with the tubular members erected plate by plate, using 2-ton hydraulic cranes.

Tower Bridge

Although the designers of the Forth Bridge did not break any new ground—they employed tried and tested technology,

but on a massive scale—its appearance at the time was dramatically modern. The same cannot be said of another bridge, equally well known, that was completed five years later. Tower Bridge is a bascule, or moving bridge. There have been many such bridges over the centuries, starting with the well-known drawbridges of medieval castles, which could be raised to frustrate attackers. Even Old London Bridge incorporated a defensive rising section. Other types of moving bridge include the swing bridge, which pivots from a central pier; rolling bascules; lift bridges; and transporter bridges, which are more like airborne ferries than bridges. However, Tower Bridge is a drawbridge or, in technical terms, a double-leaf bascule with suspended side spans.

The proximity of the new bridge to the Tower of London required the designers—engineer Sir John Wolfe Barry and architect Sir Horace Jones—to give the bridge an appropriately Gothic appearance. The Gothic cladding, with its pinnacles and pointed roofs, is more French than English in style. Tower Bridge earned aesthetic scorn when it was built, but it is now such a well-loved landmark that it is impossible to imagine London without it.

RIGHT: The familiar profile of London's Tower Bridge opens to let a luxury cruise ship pass through to the upper reaches of the Thames. The process of raising and lowering the roadway of this double-leaf bascule bridge can take up to an hour.

LEFT: An unusual view of Tower Bridge, London. Unless tourists take a river trip on the Thames, they are unlikely to get this close a shot of the trussed walkways, which bear the arms of the City of London.

BELOW: Passing under Tower Bridge, London, it is easy to observe the curved trusses that support the side spans, which project incongruously from the Gothic-style masonry concealing the steel-framed towers beneath.

PERSPECTIVE VIEW AT ANCHORAGE PIERS.

Record-Breaking Steel Structures

Steel did not yet have the field to itself; some years before, Gustave Eiffel (1832–1923) had built a series of wrought-iron viaducts in the Massif Central region in France. To combat the high winds for which the area is known, he used continuous open-truss spans on splayed piers made of open ironwork, with tubular corner columns to reduce wind resistance. In 1875, he won a competition for the Pia Maria railway bridge over the Douro River near Oporto in Portugal, and in 1884 he completed his greatest bridge, the Garabit Viaduct over the Truyère River near St. Flour in the South of France.

In 1898, a new record for steel arches was set by the Niagara–Clifton Bridge in the United States, with a span of 840 feet (256 meters). Completed in the same year, the first large steel bridge in France was the Viaur Viaduct, with a central cantilever span of 721 feet (220 meters); in 1907, the English engineer Ralph Freeman built a steel arch with a 500-foot (152-meter) span across then spectacular Zambezi Gorge below the Victoria Falls in Africa. The two halves of the arch were built from each side of the gorge, held by cables until they met in the middle.

The Queensboro Bridge in New York, designed by Gustav Lindenthal (1850–1935), was the first really large cantilever bridge built in America. Unlike the Forth Bridge it has no central suspended sections; the cantilevers themselves meet over the water. The first bridge to be constructed in high-strength nickel steel, it has two unequal main spans of 1,182 feet (360 meters), with a central span of 630 feet (192 meters) and two outer spans of 469 feet (143 meters) and 459 feet (140 meters).

The most famous arch bridge in the world was completed sixteen years later, on the other side of the world. With a main span of 1,650 feet (503 meters), the Sydney Harbor Bridge was the widest-ever long-span bridge, with four rail tracks and six lanes of roadway, although the record for length had been set the year before by the 1,652-foot (504-meter) Bayonne Bridge which crosses the Kill Van Kull between Newark and Staten Island.

Nevertheless, the Sydney Harbor Bridge can justly be called the world's greatest steel arch, not only because of its immense

BELOW: A small-scale architectural model of a pier of the Queensboro Bridge over the East River, which was opened in 1909, connecting the boroughs of Manhattan and Queens (which formally entered Greater New York in 1878).

LEFT: This perspective view of the pier, roadway, and subway tracks designed for the Queensboro Bridge at the beginning of the twentieth century is an elegant testimonial to the solid construction of the bridge, which still serves millions of travelers more than ninety years after it was designed.

FOLLOWING PAGE:
The scenic Sylvenstein Dam Bridge in Bavaria. This beautiful area features rich, softly rolling hills, and is bordered by mountain ranges.

feet (335 meters) each. Each suspended section weighs 650 tons; they were lifted into place using huge sand-filled boxes as counterweights in about half an hour (compared with the four days taken for the Quebec Bridge opened in 1919).

However, the largest cantilever bridge to be completed at that time was built in India by British engineers. The firm of Rendel, Palmer and Tritton designed the 1,500-foot (457-meter) span of the Howrah River Bridge, which crosses the Hooghly River at Calcutta. It was intended to replace a floating timber pontoon bridge that had been installed in 1874, and was designed to last only until the turn of the century. The piers of the new Howrah Bridge were the largest ever sunk at the time, each containing more than 40,000 tons of concrete, and bedded on foundations 100 feet (30 meters) deep.

carrying capacity, but also because of the tremendous difficulties that had to be overcome in erecting it across a deep harbor in which no temporary supports were possible. The designer was Ralph Freeman of the English company Dorman Long. Most of the 38,390-ton arch is composed of high-tensile silicon steel, made in Britain and fabricated in purpose-built shops in Sydney. The two halves of the arch were built out as cantilevers, held back by wire ropes until they met. Two cranes, traveling along the upper chord (curve) of the arch, were used to assemble the steelwork.

In 1938, the first Carquinez Strait Bridge at San Francisco pushed the frontiers of cantilever construction technology still further. It was designed by David Steinman (1887–1960), who had at one time been an assistant to Gustav Lindenthal. It is a double cantilever, comparable in size to the Queensboro Bridge, with two main spans of 1,100

Concrete

The Romans' discovery of cement had been one of the key factors in their development of bridge-building; but that knowledge died with the fall of the Roman Empire. It was not until the late eighteenth century that John Smeaton developed new waterproof pozzolanic cements. In 1824, Joseph Aspdin patented the process producing the new artificial Portland cement, which was soon being used for pointing and facing masonry and for the foundations of civil engineering works.

Concrete is strong in compression but, like stone, it is weak in tension. In reinforced concrete, the compressive strength is combined with the tensile strength of steel. In the early nineteenth century, various attempts had been made to strengthen

ABOVE: Although Australia's Sydney Harbor Bridge is not the longest arch bridge in the world, it is probably the most famous. Just visible beyond the bridge is a more recent engineering marvel—the dramatic roof line of the Sydney Opera House.

LEFT: The Quebec Bridge across the St. Lawrence River in Canada is the world's longest cantilever span. Work began on the bridge in 1899, and it was finally opened on December 3, 1917, at a cost of millions of dollars and eighty-six lives.

concrete by embedding iron in it. However, it is generally accepted that the modern use of reinforced concrete has its origins in a patent taken out in 1867 by a French gardener named Joseph Monier, for making plant tubs out of cement mortar strengthened with embedded iron netting.

Concrete was regarded as a replacement for stone, and so most of the early concrete bridges were built in the traditional arch form. The first reinforced-concrete bridge was built in 1869; it was a 44-foot (13-meter) shallow-arch pedestrian bridge in the castle park of the French Marquis Tilière de Chazelet. The first major use of concrete in Britain was not seen until the Glenfinnan Railway Viaduct in Invernessshire, Scotland, was completed in 1898.

In the 1880s, François Hennebique in France researched the design of beams, substituting steel for iron, and paralleling Hyatt's earlier work. In Germany, G. A. Wayss had bought the rights to Monier's patent; by the closing years of the century his company, Wayss and Freitag, had built a number of heavy arch bridges, with spans of up to 100 feet (30 meters). Hennebique had patented his ideas in 1892, and by 1900 over a hundred bridges had been built to his designs.

Born in 1872, five years after Monier took out his original patent, the Swiss engineer Robert Maillart was to become

arguably the most influential figure in the design of concrete bridges. His first works—the Stauffacher arch over the Sihl River in Zurich, and bridges at Zuoz and Billwil—were relatively traditional in form. But he broke new ground with the Tavanasa Bridge over the Rhine in 1905. To avoid the cracking that had appeared in the walls of the Zuoz bridge, he left these areas open in the form of triangular cutouts, with the span resting on slender fingers of concrete. Maillart died in 1940, but what is probably his best-known work was completed in 1930: the Salginatobel Bridge near Schiers soars across a gorge in the Graubünden, 250 feet (76 meters) above the valley floor beneath.

Plougastel Bridge

Contemporary with the Salginatobel Bridge, but far grander in scale, is the Plougastel Bridge in Brittany, France. This was designed by Eugène Freyssinet (1879–1962), who ranks alongside Maillart as one of the pioneers of modern concrete bridge-building. He had demonstrated the cost advantages of the new material early in his career, when in 1907 he built three bridges across the River Allier for less than the cost of one traditional stone bridge. The Veurdre, Boutiron, and Châtel-de-Neuvre bridges each had three spans of more than 230 feet (70 meters).

ABOVE: Looking upward at the Eisenbahn Bridge, which spans the Tal near Schlitz in Hesse, one can see the power inherent in this form, which reaches to the sky from the earth below.

RIGHT: The Tauern Motorway, in Germany's Lieger Valley, is a splendid example of the efficient means of transportation available on some 6,500 miles (10,500 kilometers) of toll-free highways.

LEFT: Built using traveling formwork, the Kochertal Bridge in Geislingen (Germany) spans 3,700 feet (1,128 meters). There are two launching girders, which advance from both abutments toward the center; other loads are carried by the cantilevering concrete girder.

BELOW: The Zeeland Bridge, a traffic bridge over the Oosterschelde in the Netherlands, is part of the highway network between the Dutch Randstad and the southwestern delta area. Because the water currents and weather in the Oosterschelde create extremely difficult working conditions, prefabrication was used to a large extent in the construction process.

As a result of a full-scale prototype test, Freyssinet had discovered the phenomenon of concrete creep, which occurs as the material continues to shrink slightly after it has solidified. He was able to allow for this on the Veurdre Bridge by leaving a slight gap at the crown of each arch, which he then jacked apart and filled with additional concrete a year later.

Freyssinet completed the Plougastel Bridge in 1930, construction having started in 1925. It consists of three segmental arch spans, each 592 feet (180 meters), across the Elorn estuary. The hollow, concrete-box arches are 31 feet (9.5 meters) wide and 14 feet, 8 inches (4.5 meters) deep at the centers, 90 feet (27.5 meters) above the water. As a result of his studies on the creep phenomenon at Plougastel, Freyssinet came to the conclusion that, in effect, a new material could be created by prestressing with high-strength steel bonded to the concrete. Prestressing is rather like stretching a single rope taut: longitudinal steel strands in a concrete beam are stretched and then anchored to the ends of the beam. This increases the compressive strength of the concrete far more effectively than simple reinforcement.

The first prestressed-concrete bridge was the Saale Bridge in Alsleben, built in 1928 to a design by Franz Dischinger. It had an arch span of 223 feet (68 meters), with a steel tension ribbon in the arch. One of the longest precast, prestressed-concrete beam bridges in the world is in India: the Ganga Bridge at Patna has forty-five cantilevered main spans, each of 397 feet (121 meters). It was opened when only side had been built, but was finally completed in 1982.

ABOVE: A bridge is one way to cross the St. Johns River in Jacksonville, the original gateway to Florida's famous beach resort areas.

RIGHT: The modern Seven-Mile Bridge (a part of the Overseas Highway, U.S. 1), in the Florida Keys, was built on top of a branch of the Florida East Coast Railroad created by oil magnate Henry Flagler, which operated between 1912 and 1935, until a portion of it was destroyed during a hurricane.

CHAPTER THREE

SUSPENSION AND CABLE-STAYED BRIDGES

Every structural material has a specific ratio of strength to weight. For modern high-tensile steel, the theoretical maximum span for a cantilever bridge is about 2,500 feet (760 meters); for arch bridges it is about 300 feet (90 meters). Steel-wire cables have a much higher strength-to-weight ratio than structural steel, and hence suspension bridges can be built with much longer spans. Suspension bridges, such as the Golden Gate Bridge in San Francisco and the Humber Estuary Bridge in England, are often thought of as epitomizing modern bridge design. But the suspension principle is a very old idea. In its simplest form, it is capable of spanning much

greater distances with far fewer engineering resources than any other type of structure.

A single rope or cable hanging across a canyon is the most primitive form of suspension bridge. Further sophistication is added with a bottom and sides woven of many cables, or supporting short planks of wood. Iron chain suspension structures were found in China in the eighth century, and there was even a chain-suspended bridge in early medieval Europe. This was the Twärrenbrücke, or Transverse Bridge, erected in about 1218 across the Schöllene Canyon in what is now Switzerland.

ABOVE: The Barqueta Bridge at Expo '92 in Seville, Spain, was designed by J. J. Arenas and M. J. Pantaleon. Its combination of arch and cable structure typifies the exciting new hybrid designs being produced by today's engineers.

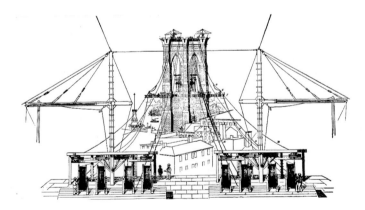

ABOVE: This drawing shows the bank of cable-spinning machines used in the construction of New York's Brooklyn Bridge, a marvel of engineering innovation in its time.

RIGHT: The George Washington Bridge over the Hudson River in New York City was designed by the Swiss-born engineer Othmar Ammann.

The distinctive truss structures of the towers, shown here, were originally intended to be covered in concrete and then clad in granite.

In 1801, an American judge and justice of the peace, James Finley, built a 69-foot (21-meter) span across Jacob's Creek in Payette County, Pennsylvania, in imitation of an early-seventeenth-century design. This was the first suspension bridge to have a horizontal deck that was braced by trusses (a truss is a frame of members in tension and compression).

Finley's ideas were first developed in Britain. In 1820, Captain Samuel Brown constructed the Union Bridge across the Tweed, linking England and Scotland. With a span of 449 feet (137 meters), it was the earliest major suspension bridge. The deck was supported by vertical rods from three pairs of wrought-iron chains, which are still intact today, although supplemented by modern steel cables.

Meanwhile, Thomas Telford had been commissioned to design a bridge across the Menai Strait between the mainland of Wales and the island of Anglesey. The suspension design that he produced was truly visionary. The main span, at 579 feet (176 meters), was much longer than Brown's entire Union Bridge, and the chains were to hang from towering masonry approach viaducts.

The bridge was completed in 1826, but the specter of poor wind resistance, which was to haunt suspension-bridge design for many years, soon made its appearance. Before the bridge opened the deck had shown signs of flexing, and subsequent gales in 1826 had similar effects. The deck was strengthened, but in 1839 severe storms broke several hangers and wrecked the deck. It was rebuilt, but continued to suffer wind damage. It was rebuilt yet again in 1892, and between 1938 and 1941, when the iron chains were replaced with steel.

BELOW: The main span of Hamburg's Köhlbrand Bridge, built using traveling formwork, measures 1,066 feet (325 meters). The deck of the bridge is 197 feet (60 meters) above the water; the approach spans 230 feet (70 meters).

Wire Cables

The next major development in the design of suspension bridges was the substitution of wire cables for chains. A little-known engineer named Joseph Chaley designed the next record-breaking suspension bridge—the Grand Pont Suspendu, built over the Sarine Valley at Fribourg in Switzerland in 1834. It was 896 feet (273 meters) long, half as long again as the Menai Bridge, and is still the longest single span in Switzerland. The bridge was far too long for the cables to be prefabricated in one piece, so Chaley developed a method of "spinning" the cables together in the air from over a thousand wires grouped in twenty strands. This new technique was subsequently to be developed on the other side of the Atlantic, where the impetus for advances in suspension technology was to remain for more than a century.

In England, Isambard Kingdom Brunel's final work, completed after his death, was the Clifton Suspension Bridge across the Avon Gorge at Bristol. Despite a suggestion that he should suspend the deck from wire cables, as in Chaley's bridge, Brunel preferred to follow Telford's design and use chains. The bridge

was a long time in construction. Work began in 1831, but was repeatedly delayed by lack of money. Finally in 1842 work ceased, and the iron chains were sold. After Brunel's death in 1859, fellow members of the Institution of Civil Engineers formed a company to complete the bridge. They were able to reuse the chains from Brunel's 1845 Hungerford Suspension Bridge, which was then being demolished, and the Clifton Suspension Bridge was opened in 1864, spanning an impressive 702 feet (214 meters) across the Avon Gorge.

Charles Ellet (1810–1862) was the first American engineer to explore James Finley's ideas. His greatest work was completed in 1849. With a span of 1,010 feet (308 meters), the Wheeling Bridge over the Ohio River shattered the record set by Chaley's Grand Pont Suspendu. Despite Ellet's claims that a strongly framed deck floor would offer adequate resistance to

ABOVE: Completed in 1855, the Niagara Bridge was designed by John Roebling, a stout defender of the superiority of suspension cables over chains. This ambitious structure over Niagara Gorge featured a double deck, with the roadway running below the railroad.

RIGHT: The Ganter Bridge over the Simplon Pass in the Valais was designed by the Swiss engineer Christian Menn. This cable-stayed bridge has a straight main span of 571 feet (174 meters) and two curved side spans, each 417 feet (127 meters). The cable stats are enclosed in concrete to enable them to follow the curved plan.

An arch bridge over Roosevelt Lake, one of Arizona's prime recreation areas for camping, waterskiing, and boating. The largest of the four Salt River lakes, Roosevelt Lake stretches twenty-three miles (37 kilometers) and is, at points, nearly two miles wide (3.2 kilometers).

high winds, the bridge deck was blown down by a violent gale in 1854, and subsequently rebuilt.

In 1847 Ellet was invited to build a suspension bridge across Niagara Gorge, a prospect that had fired his enthusiasm since 1830. But he resigned after a disagreement with the bridge company, which instead called in John Roebling (1806–1869), who was a stout defender of the superiority of suspension cables over chains. Roebling had already patented a technique for compacting the wires into cylinders and wrapping them with more wire for weather protection.

The span of the Niagara Bridge was less than that of either the Grand Pont Suspendu or the Wheeling Bridge, but as a structure it was far more ambitious. Completed in 1855, it featured a double deck, with the roadway running below the rail deck, and separated from it by vertical wooden members to form a huge trussed girder. Roebling was very conscious of the dangers of oscillations caused by the wind, which, he wrote, "will increase to a certain extent by their own effect, until . . . a momentum of force may be produced that may prove stronger than the cables," prophetic words that were to be proved correct nearly a hundred years later.

In December 1966, Roebling's bridge across the Ohio River at Cincinnati was completed, with a world-record-breaking

ABOVE: The world's highest bridge is the Royal George over the Arkansas River in Colorado. A suspension bridge, it has a main span of 880 feet (268 meters) and stands 1,053 feet (321 meters) above the water level. It was built in a mere three months, and was opened on December 6, 1929.

LEFT: The first Bosporus Bridge at Istanbul connects the continents of Asia and Europe. Note the slender airfoil deck design. When the bridge was built in 1973 its span, at 3,523 feet (1,074 meters), was the longest in Europe and the fourth longest in the world.

LEFT: The suspension bridge over the River Tagus, near Lisbon, Portugal, was opened in 1966. The high deck clearance of 230 feet (70 meters) was required to allow shipping free passage beneath. The deep truss was designed to accommodate a lower rail deck, but this has never been added.

LEFT: The Severin Bridge in Cologne was built in 1960. This cable-stayed bridge has an asymmetrically placed pylon— deliberately designed in this fashion because two pylons would have impaired the view of Cologne cathedral from upriver.

FOLLOWING PAGE:
When it was built, the deck of the Golden Gate Bridge was shallower than in any other suspension bridge to date. The stiffening truss, shown here, later required an additional 4,700 tons (4,230 metric tons) of lateral bracing to counteract the rippling effect of cross-winds.

ABOVE: The Elizabeth Bridge over the River Danube, Budapest. When built in 1903, it was the longest eyebar chain suspension span, at 950 feet (290 meters). As dusk falls, lighting emphasizes the contrast between the elegance of the suspended chains and the massiveness of the supporting towers.

RIGHT: The Albert Bridge, over the River Thames at Battersea, London, is essentially a cable-stayed design. Designed by R. M. Ordish, it was built between 1871 and 1873. The main span is 400 feet (122 meters).

LEFT: The Pont de Normandie, across the Seine Estuary near Le Havre in France. This photograph, taken in 1994, shows the unfinished span hanging over the Seine Estuary like a diver's springboard.

span of 1,057 feet (322 meters). He then went on to conceive the crowning achievement of his career, but one that, sadly, he was not to see completed. His design proposal for a 1,500-foot (457-meter) bridge across New York's East River, to link Brooklyn and Manhattan Island, was masterly in its scope and detail; but tragically he was killed in an accident while surveying the site and his son, Colonel Washington Roebling (1837–1926), took over the project.

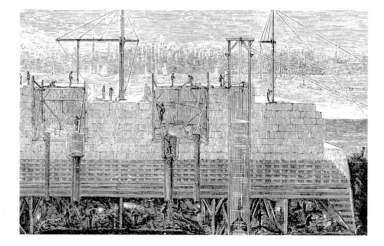

ABOVE: This contemporary engraving shows excavation work in progress for the foundations of Brooklyn Bridge. The work was immensely difficult and very dangerous.

Long-Span Classics

The Brooklyn Bridge was the first long-span bridge in New York, but twenty years later it was joined by the Williamsburg Bridge, spanning 1,600 feet (488 meters) over the East River just to the northeast. This was the first large suspension bridge to have steel towers. So too did the Manhattan Bridge, built just three years later in 1906. Now New York had three crossings over the East River, all within less than 2 miles (3 kilometers) of each other. The Manhattan Bridge had been designed by Gustav Lindenthal, who had supervised the completion of the Williamsburg Bridge. In 1923 his assistant, Othmar Amman, put forward a proposal for a bridge upstream of Lindenthal's. The George Washington Bridge was begun in 1927 and completed in 1931; it was truly state-of-the-art construction. The span, at 3,500 feet (1,067 meters), was nearly double the previous unsupported span. The four suspension cables contained a total of some 107,000 miles (173,000 kilometers) of wire!

If the various New York bridges are symbols of that city, how much more so is the Golden Gate Bridge in San Francisco. It was designed by Charles Ellis, with the red lead color and unique detailing provided by the architect Irving Morrow. The design brief was challenging. Structural safety was paramount, with the potential for both gales and earthquakes, but the deck had to be particularly high to allow ships to pass beneath, and the beauty of the site demanded a design to match.

LEFT: In this dramatic close-up view of the Brooklyn Bridge, the radiating stays and vertical hangers criss-cross like threads in a spider's web. At the time of its completion, this was the world's longest suspension bridge.

RIGHT: This shot of the Brooklyn Bridge in New York clearly shows the intricate network of stays, their slenderness contrasting vividly with the girth of the main cables and the bulk of the supporting towers.

RIGHT: Once the cables of the Brooklyn Bridge were in place, construction of the bridge floor could begin. This photograph, taken in 1881, shows work in progress. The bridge, designed by J. A. Roebling and his son, W. A. Roebling, was built between 1869 and 1883.

RIGHT: Brooklyn Bridge under construction in 1872. The men working in the pneumatic caissons beneath this growing mass of masonry lived in constant fear of being crushed if the supporting timber platform were to collapse.

LEFT: In the twentieth century, suspension bridge design has become ever more slender and elegant. The towers of the Golden Gate Bridge, San Francisco, are a world away from the masonry of New York's Brooklyn Bridge. The bridge was designed by Charles Ellis, but the red lead color and unique detailing were contributed by the architect Irving Morrow.

ABOVE: This famous photograph was taken as a large section of the concrete roadway in the center span of the Tacoma Narrows Bridge (Washington State) hurtled into Puget Sound on November 7, 1940. High winds caused the bridge to sway, undulate, and finally collapse under the strain.

BELOW: Architectural drawing for the arch and colonnade of the Manhattan approach to the Manhattan Bridge. This classic suspension bridge, which spans the boroughs of Manhattan and Brooklyn, was designed by the architectural firm of Carrère and Hastings, and constructed between 1904 and 1909.

At 4,200 feet (1,280 meters), the suspended deck is 700 feet (213 meters) longer than that of the George Washington Bridge, but it was narrower, and the most slender in relation to the span than anything yet built—so much so that additional stiffening had subsequently to be added. With this addition, the Golden Gate Bridge has stood, uniquely distinctive, ever since.

The Tacoma Narrows Bridge, over Puget Sound in Washington State, stood for a mere four months. Designed by Leon Moisseiff, it was the most slender suspension bridge yet built. Its suspended span was 2,800 feet (853 meters), but the plate girder that supported the deck was a mere 8 feet (2.4 meters) thick, with precious little in the way of cross-bracing to provide any additional stiffness. As soon as the bridge was opened in July 1940, it was nicknamed "Galloping Gertie" because of the way even a light wind made the deck sway from side to side and—more alarmingly—set up torsional (twisting) vibrations that rippled along the length of the span. Finally, in a wind of only 42 miles per hour (68 kilmeters per hour), the vibrations became so violent that the deck was torn away and crashed into the water. Other American bridges had also shown undesirable action, and further bracing or stiffening was quickly added.

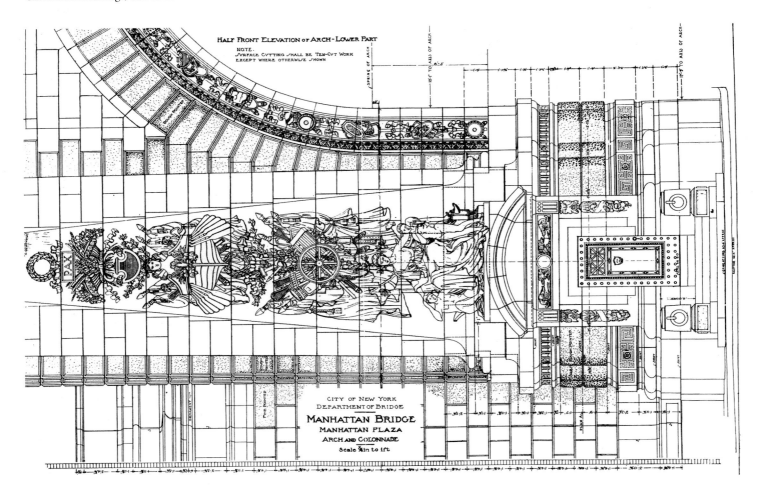

European Developments

This approach was adopted in Britain in the new Forth Road Bridge (1964), the first European bridge to rival achievements in America. Its design followed broadly similar principles to those of recent American examples, but the steel trusswork was more sparing and delicate. At 3,300 feet (1,006 meters), the main span was three-quarters the length of the Verrazano Narrows Bridge, but also almost three-quarters lighter.

If the Forth Road Bridge modified the American tradition of bridge-building, the Severn Bridge linking England and Wales departed from it altogether, and saw the impetus for developing the suspension form move back across the Atlantic. Prompted by the ideas of Fritz Leonhardt, a renowned European bridge engineer, the designers of the Severn Bridge adopted a very different solution to the problem of the aerodynamic forces on a bridge deck. Rather than stiffening it with trusses, they adopted design principles

ABOVE: New York's East River has three suspension crossings, all within 2 miles (3.2 kilometers) of each other. This picture shows John Roebling's Brooklyn Bridge, completed in 1883, and beyond it the Manhattan Bridge, designed by Gustav Lindenthal and completed in 1909.

similar to those that aircraft designers had been using for many years. The deck was designed as a box girder shaped like an airfoil. The roadway was constructed on the upper surface of the deck, which is only 10 feet (3 meters) deep, and is suspended from the cables by inclined hangers at 60-foot (18-meter) intervals.

This design solution, elegant in every way, results in a considerable weight saving, both in the deck itself, and in the cables and towers, which can be lighter because of the reduced wind loads on the bridge. The Severn Bridge has been described as "the first suspension bridge of the modern type."

Cable-Stayed Bridges

In a cable-stayed bridge, the deck is stabilized and support-ed by ropes from a vertical support. The idea is not new; it could be said to have begun with the booms, rigging, and masts of Egyptian sailing ships.

The bridges that were built in the nineteenth century com-bined radiating stays with catenary suspension cables as in the Franz Joseph Bridge in Prague, opened in 1868. The bridge that was closest to modern designs was a small concrete aque-duct built by Torroja over the Guadalete River in Spain in 1925. In 1938, Dischinger realized that he could achieve stiff-ness and stability if the cables were made of high-strength wires under stress, and Fritz Leonhardt was developing "orthotropic" steel decks, with girders along the underside of the deck plates. These were the two elements that made possi-ble the explosion of cable-stayed designs after World War II.

Completed in 1962, Riccardo Morandi's design for the Lake Maracaibo Bridge in Venezuela broke new ground. Built in rein-forced and prestressed concrete, it features a central cable-stayed span that is 1,300 feet (396 meters) long. Early bridges such as

LEFT: Aircraft vapor trails echo the curves of the cables in this dramatic view from one of the towers of the Verrazano Narrows Bridge across New York harbor. When it was opened in 1964 it was the biggest bridge in history; the suspended span is 4,260 feet (298 meters), each tower contains 27,000 tons (24,300 metric tons) of steel, and the main cables contain enough wire to go around the earth five times.

ABOVE: Completed in 1939, the Lion's Gate Bridge spans 1,550 feet (472 meters) across Canada's Vancouver harbor, one of the many possible routes in this busy Pacific port's extension transportation network.

LEFT: The Mackinack Straits Bridge in Michigan was engineer David Steinman's greatest design. With a suspended span of 3,800 feet (1,158 meters) and a total span of 8,614 feet (2,626 meters), it contains some 55,000 tons (49,500 metric tons) of structural steel. The cables alone weigh 11,000 tons (9, 900 metric tons).

ABOVE: The Glebe Island Bridge in Sydney, the longest cable-stayed bridge in Australia, is a 2,640-foot (805-meter) concrete structure comprising six spans with three elegant cable-stayed central spans.

this had few and relatively massive stays; this required equally substantial stays in the deck, which also had to be correspondingly deep. Subsequent bridges had more numerous, finer supports, which meant that the depth of the deck could be reduced.

The Annacis Bridge (now the Alex Fraser Bridge) over the Fraser River near Vancouver was the first cable-stayed bridge to be built of composite construction. In such a bridge, a thin reinforced or prestressed concrete deck is supported by a frame of steel girders. In the Annacis Bridge, the concrete panels are only 8.7 inches (223 millimeters) thick. They are supported by a steel frame, which was built out from either side of the 596-foot-high (182-meter) concrete towers as a balanced cantilever.

The main span of the Annacis Bridge is 1,526 feet (465 meters). This record width was not exceeded until 1991, by the 1,739-foot (530-meter) Skarnsundet Bridge in Norway. For its length, the bridge is remarkably narrow: only 7 feet (2.1 meters) deep and 43 feet (13 meters) wide. To ensure adequate stiffness, the deck has a hollow, triangular section. The cables are fixed to A-shaped concrete pylons 499 feet (152 meters) above water level.

Spanning the Future

Today, records are constantly being broken as engineers push materials ever nearer their limits and devise new methods of construction. But for sheer length, it is the suspension bridges that will always dominate. In 1980, the record-breaking Humber Bridge was opened, with a span of 4,624 feet (1,409 meters) across the Humber Estuary on the northeast coast of England. The main towers are over 533 feet (162 meters) tall, and are 1 3/8 inches (36 millimeters) out of parallel—to allow for the curvature of the earth!

When the Akashi–Kaikyo Bridge is completed in 1998, to link the islands of Honshu and Shikoku, its main span will be 6,528 feet (1,990 meters) long. Plans are afoot to build a suspension bridge across the Strait of Messina, in Italy, that

ABOVE: The Waal Bridge, Zaltbommel (Netherlands) transverses the busiest river in Europe, and consitutes the main transport connection between Rotterdam harbor and the German hinterland. This ultra-modern cable-stayed bridge began construction in September of 1992 and was completed in January of 1996.

LEFT: A modern highway bridge in Sluis offers access to this historic fortified port, which is today a lively tourist center in the Netherlands.

FOLLOWING PAGE:
The Seto Ohashi Bridge, near Okayama (Japan), links Shikoku and Honshu Island. It is the world's longest double-deck bridge—an important link for this industrial center situated on an inlet of the Inland Sea.

would be about 2 miles (3.3 kilometers) between the main towers. And the visionary engineer T. Y. Lin has proposed a series of cable-stayed bridges that would join North America and Russia across the Bering Strait—an Intercontinental Peace Bridge.

So, it seems, as our understanding of forces and structures deepens, and the range of available materials increases, so

BELOW: A modern highway bridge spans the Fehmarn Sound in the state of Schleswig-Holstein, linking the island of Fehmarn to the German mainland.

bridge designs become more daring, more ambitious. Spans are ever longer, structures ever more lightweight—is there a limit? For any given material, the answer is a resounding "yes." A bridge has to be able to support its own weight without collapsing—this requirement defines its limiting span. The thought of a single steel arch, for example, that is nearly a mile across is a daunting prospect. But other, newer materials offer tantalizing possibilities of even greater spans.

The Aberfeldy Golf Club in Scotland recently commissioned a footbridge, to span 207 feet (63 meters)—a modest structure, by today's standards. But the bridge was commissioned from a company called Maunsell Structural Plastics, and it is to be built entirely in composites—plastic reinforced with glass and cables of Kevlar in a polyethylene coating. Kevlar, an aramid fiber, is one of the new generation of bridge-building materials. Another is carbon fiber, which, weight for weight, has four times the strength of high-tensile steel. It has been estimated that the limiting span for a carbon or Kevlar suspension span would be 7.5 miles (12 kilometers)—more than eight times the span of the Humber Bridge.

In a world torn apart by war, poverty, and disease, such talk can be seen as empty vanity. But not for nothing do we talk of "building bridges" when we try to cross social, cultural, and political divides. This book has concentrated on bridges as engineering constructions, but behind and beyond the mere technology, bridges are metaphors of the inextinguishable human spirit. As long as there are people on this planet, they will be building bridges.

ABOVE: During the 1960s, the construction of the MostSNP Bridge over the Danube River in Bratislava (CSR) destroyed some of the old buildings from this historic city, which was the capital of Hungary for over two hundred years.

ABOVE: The Köhlbrand Bridge in Hamburg was built in 1974. The base of the A-frame tower curves in on itself to enclose the deck. The approach viaducts, box-girder deck, and main piers are concrete; the towers and central span are steel.

LEFT: The Sunshine Skyway Bridge over Tampa Bay, Florida is the largest concrete span in North America. Dramatically silhouetted against the evening sky, the cable-stayed structure seems to sail across the bay like a giant version of the yacht passing beneath.

LEFT: The supporting cables of the Sunshine Skyway Bridge in Tampa Bay, Florida, can take on a form reminiscent of an abstract sculpture.

ABOVE: Still under construction in 1996, Rotterdam's ultra-modern
Erasmus Bridge, which spans the Maas, is named after the
famous humanist and theologian Desiderius Erasmus (1469–1536).

RIGHT: The E-4 Bridge crosses from the mainland
to the small island of Bogø, in Denmark. More
distant islands are served by an elaborate ferry system.

INDEX